The World With A Thousand Moons

by

Edmond Hamilton

ABOUT THE AUTHOR

Edmond Moore Hamilton (October 21, 1904 – February 1, 1977) was a mid-twentieth-century American science fiction writer. He was born in Youngstown, Ohio, and raised in adjacent New Castle, Pennsylvania. He graduated from high school and enrolled in Westminster College in New Wilmington, Pennsylvania, at the age of 14, but dropped out when he was 17. Edmond Hamilton's career as a science fiction writer began with the publication of his short tale "The Monster God of Mamurth" in the August 1926 edition of Weird Tales, which is today considered a classic journal of alternative fiction. Hamilton rapidly established himself as a key member of Farnsworth Wright's amazing circle of Weird Tales writers, which included H. P. Lovecraft and Robert E. Howard. From 1926 to 1948, Weird Tales published 79 works of fiction by Hamilton, making him one of the magazine's most prolific contributors. Hamilton became friends and associates with several Weird Tales veterans, including E. Hoffmann Price and Otis Adelbert Kline; most famously, he had a 20-year connection with near contemporary Jack Williamson, as Williamson recounts in his 1984 autobiography Wonder's Child.

CONTENTS

CHAPTER I
THRILL CRUISE

Lance Kenniston felt the cold realization of failure as he came out of the building into the sharp chill of the Martian night. He stood for a moment, his lean, drawn face haggard in the light of the two hurtling moons.

He looked hopelessly across the dark spaceport. It was a large one, for this ancient town of Syrtis was the main port of Mars. The forked light of the flying moons showed many ships docked on the tarmac—a big liner, several freighters, a small, shining cruiser and other small craft. And for lack of one of those ships, his hopes were ruined!

A squat, brawny figure in shapeless space-jacket came to Kenniston's side. It was Holk Or, the Jovian who had been waiting for him.

"What luck?" asked the Jovian in a rumbling whisper.

"It's hopeless," Kenniston answered heavily. "There isn't a small cruiser to be had at any price. The meteor-miners buy up all small ships here."

"The devil!" muttered Holk Or, dismayed. "What are we going to do? Go on to Earth and get a cruiser there?"

"We can't do that," Kenniston answered. "You know we've got to get back to that asteroid within two weeks. We've got to get a ship here."

Desperation made Kenniston's voice taut. His lean, hard face was bleak with knowledge of disastrous failure.

The big Jovian scratched his head. In the shifting moonslight his battered green face expressed ignorant perplexity as he stared across the busy spaceport.

"That shiny little cruiser there would be just the thing," Holk Or muttered, looking at the gleaming, torpedo-shaped craft nearby. "It would hold all the stuff we've got to take; and with robot controls we two could run it."

"We haven't a chance to get that craft," Kenniston told him. "I found out that it's under charter to a bunch of rich Earth youngsters who came out here in it for a pleasure cruise. A girl named Loring, heiress to Loring Radium, is the head of the party."

The Jovian swore. "Just the ship we need, and a lot of spoiled kids are using it for thrill-hunting!"

Kenniston had an idea. "It might be," he said slowly, "that they're tired of the cruise by this time and would sell us the craft. I think I'll go up to the Terra Hotel and see this Loring girl."

"Sure, let's try it anyway," Holk Or agreed.

The Earthman looked at him anxiously. "Oughtn't you to keep under cover, Holk? The Planet Patrol has had your record on file for a long time. If you happened to be recognized—"

"Bah, they think I'm dead, don't they?" scoffed the Jovian. "There's no danger of us getting picked up."

Kenniston was not so sure, but he was too driven by urgent need to waste time in argument. With the Jovian clumping along beside him, he made his way from the spaceport across the ancient Martian city.

The dark streets of old Syrtis were not crowded. Martians are not a nocturnal people and only a few were abroad in the chill darkness, even they being wrapped in heavy synthewool cloaks from which only their bald red heads and solemn, cadaverous faces protruded.

Earthmen were fairly numerous in this main port of the planet. Swaggering space-sailors, prosperous-looking traders and rough meteor-miners made up the most of them. There were a few tourists gaping at the grotesque old black stone buildings, and under a krypton-bulb at a corner, two men in the drab uniform of the Patrol stood eyeing passersby sharply. Kenniston breathed more easily when he and the Jovian had passed the two officers without challenge.

The Terra Hotel stood in a garden at the edge of town, fronting the moonlit immensity of the desert. This glittering glass block, especially built to cater to the tourist trade from Earth, was Earth-conditioned inside. Its gravitation, air pressure and humidity were ingeniously maintained at Earth standards for the greater comfort of its patrons.

Kenniston felt oddly oppressed by the warm, soft air inside the resplendent lobby. He had spent so much of his time away from Earth that he had become more or less adapted to thinner, colder atmospheres.

"Miss Gloria Loring?" repeated the immaculate young Earthman behind the information desk. His eyes appraised Kenniston's shabby space-jacket and the hulking green Jovian. "I am afraid—"

"I'm here to see her on important business, by appointment," Kenniston snapped.

The clerk melted at once. "Oh, I see! I believe that Miss Loring's party is now in The Bridge. That's our cocktail room—top floor."

Kenniston felt badly out of place, riding up in the magnetic lift with Holk Or. The other people in the car, Earthmen and women in the shimmering synthesilks of the latest formal dress, stared at him and the Jovian as though wondering how they had ever gained admittance.

The lights, silks and perfumes made Kenniston feel even shabbier than he was. All this luxury was a far cry from the hard, dangerous life he had led for so long amid the wild asteroids and moons of the outer planets.

It was worse up in the glittering cocktail room atop the hotel. The place had glassite walls and ceiling, and was designed to give an impression of the navigating bridge of a space-ship. The orchestra played behind a phony control-board of instruments and rocket-controls. Meaningless space-charts hung on the walls for decoration. It was just the sort of pretentious sham, Kenniston thought contemptuously, to appeal to tourists.

"Some crowd!" muttered Holk Or, looking over the tables of richly dressed and jewelled people. His small eyes gleamed. "What a place to loot!"

"Shut up!" Kenniston muttered hastily. He asked a waiter for the Loring party, and was conducted to a table in a corner.

There were a half dozen people at the table, most of them young Earthmen and girls. They were drinking pink Martian desert-wine, except for one sulky-looking youngster who had stuck to Earth whisky.

One of the girls turned and looked at Kenniston with cool, insolently uninterested gaze when the waiter whispered to her politely.

"I'm Gloria Loring," she drawled. "What did you want to see me about?"

She was dark and slim, and surprisingly young. There were almost childish lines to the bare shoulders revealed by her low golden gown. Her thoroughbred grace and beauty were spoiled for Kenniston by the bored look in her clear dark eyes and the faintly disdainful droop of her mouth.

The chubby, rosy youth beside her goggled in simulated amazement and terror at the battered green Jovian behind Kenniston. He set down his glass with a theatrical gesture of horror.

"This Martian liquor has got me!" he exclaimed. "I can see a little green man!"

Holk Or started wrathfully forward. "Why, that young pup—"

Kenniston hastily restrained him with a gesture. He turned back to the table. Some of the girls were giggling.

"Be quiet, Robbie," Gloria Loring was telling the chubby young comedian. She turned her cool gaze back to Kenniston. "Well?"

"Miss Loring, I heard down at the spaceport that you are the charterer of that small cruiser, the *Sunsprite*," Kenniston explained. "I need a craft like that very badly. If you would part with her, I'd be glad to pay almost any price for your charter."

The girl looked at him in astonishment. "Why in the world should I let you have our cruiser?"

Kenniston said earnestly, "Your party could travel just as well and a lot more comfortably by liner. And getting a cruiser like that is a life-or-death business for me right now."

"I'm not interested in your business, Mr. Kenniston," drawled Gloria Loring. "And I certainly don't propose to alter our plans just to help a stranger out of his difficulties."

Kenniston flushed from the cool rebuke. He stood there, suddenly feeling a savage dislike for the whole pampered group of them.

"Beside that," the girl continued, "we chose the cruiser for this trip because we wanted to get off the beaten track of liner routes, and see something new. We're going from here out to Jupiter's moons."

Kenniston perceived that these bored, spoiled youngsters were out here hunting for new thrills on the interplanetary frontier. His dislike of them increased.

A clean-cut, sober-faced young man who seemed older and more serious than the rest of the party, was speaking to the heiress.

"Unhardened space-travellers like us are likely to get hit by gravitation paralysis out in the outer planets, Gloria," he was saying to the heiress. "I don't think we ought to go farther out than Mars."

Gloria looked at him mockingly. "If you're scared, Hugh, why did you leave your nice safe office on Earth and come along with us?"

The chubby youth called Robbie laughed loudly. "We all know why Hugh Murdock came along. It's not thrills he wants—it's you, Gloria."

They were all ignoring Kenniston now. He felt that he had been dismissed but he was desperately reluctant to lose his last hope of getting a ship. Somehow he *must* get that cruiser!

A stratagem occurred to him. If these spoiled scions wouldn't give up their ship, at least he might induce them to go where he wanted.

Kenniston hesitated. It would mean leading them all into the deadliest kind of peril. But a man's life depended on it. A man who was worth all these rich young wastrels put together. He decided to try it.

"Miss Loring, if it's thrills you're after, maybe I can furnish them," Kenniston said. "Maybe we can team up on this. How would you like to go on a voyage after the biggest treasure in the System?"

"Treasure?" exclaimed the heiress surprisedly. "Where is it?"

They were all leaning forward, with quick interest. Kenniston saw that his bait had caught them.

"You've heard of John Dark, the notorious space-pirate?" he asked.

Gloria nodded. "Of course. The telenews was full of his exploits until the Patrol caught and destroyed his ship a few weeks ago."

Kenniston corrected her. "The Patrol caught up to John Dark's ship in the asteroid, but didn't completely destroy it. They gunned the pirate craft to a wreck in a running fight. But Dark's wrecked ship drifted into a dangerous zone of meteor swarms where they couldn't follow."

"I remember now—that's what the telenews said," conceded the heiress. "But Dark and his crew were undoubtedly killed, they said."

"John Dark," Kenniston went on, "looted scores of ships during his career. He amassed a hoard of jewels and precious metals. And he kept it right with him in his ship. That treasure's still in that lost wreck."

"How do you know?" asked Hugh Murdock bluntly.

"Because I found the lost wreck of Dark's ship myself," Kenniston answered. He hated to lie like this, but knew that he had no choice.

He plunged on. "I'm a meteor-miner by profession. Two weeks ago my Jovian partner and I were prospecting in the outer asteroid zone in our little rocket. Our air-tanks got low and to replenish them, we landed on the asteroid Vesta. That's the big asteroid they call the World with a Thousand Moons, because it's circled by a swarm of hundreds of meteors.

"It's a weird, jungled little world, inhabited by some very queer forms of life. In landing, my partner and I noticed where some great object had crashed down into the jungle. We discovered it was the wreck of John Dark's ship. The wreck had drifted until it crashed on Vesta, almost completely burying itself in the ground. No one was alive on it, of course."

Kenniston concluded. "We knew Dark's treasure must still be in the buried wreck. But it would take machinery and equipment to dig out the wreck. So we came here to Mars, intending to get a small cruiser, load it with the necessary equipment, and go back to Vesta and lift the treasure. Only we haven't been able to get a ship of any kind."

He leaned toward the girl. "Here's my proposition, Miss Loring. You take us and our equipment to Vesta in your cruiser, and we'll share the treasure with you fifty-fifty. What do you say?"

The blonde girl beside Gloria uttered a squeal of excitement. "Pirate treasure! Gloria, let's do it—what a thrill it would be!"

The others showed equal excitement. The romance of a treasure hunt in the wild asteroids lured them, rather than the possible rewards.

"We'd certainly be able to take back a wonderful story to Earth if we found John Dark's treasure," admitted Gloria, with quick, eager interest.

Hugh Murdock was an exception to the general enthusiasm. He asked Kenniston, "How do you know the treasure's still in the buried wreck?"

"Because the wreck was still undisturbed," Kenniston answered. "And because we found these jewels on the body of one of John Dark's crew, who had been flung clear somehow when the wreck crashed."

He held out a half-dozen gems he took from his pocket. They were Saturnian moon-stones, softly shining white jewels whose brilliance waxed and waned in perfect periodic rhythm.

"These jewels," Kenniston said, "must have been that pirate's share of the loot. You can imagine how rich John Dark's own hoard must be."

The jewels, worth many thousands, swept away the lingering incredulity of the others as Kenniston had known they would.

"You're sure no one else knows the wreck is there?" Gloria asked breathlessly.

"We kept our find absolutely secret," Kenniston told her. "But since I can't get a ship any other way, I'm willing to share the hoard with you. If I wait too long, someone else may find the wreck."

"I accept your proposition, Mr. Kenniston!" Gloria declared. "We'll start for Vesta just as soon as you can get the equipment you'll need loaded on the *Sunsprite*."

"Gloria, you're being too hasty," protested Hugh Murdock. "I've heard of this world with a Thousand Moons. There're stories of queer, unhuman creatures they call Vestans, who infest that asteroid. The danger—"

Gloria impatiently dismissed his objections. "Hugh, if you are going to start worrying about dangers again, you'd better go back to Earth and safety."

Murdock flushed and was silent. Kenniston felt a certain sympathy for the young businessman. He knew, if these others did not, just how real was the alien menace of those strange creatures, the Vestans.

"I'll go right down to the spaceport and see about loading the equipment aboard your cruiser," Kenniston told the heiress. "You'd better give me a note to your captain. We ought to be able to start tomorrow."

"Pirate treasure on an unexplored asteroid!" exulted the enthusiastic Robbie. "Ho for the World with a Thousand Moons!"

Kenniston felt guilty when he and Holk Or left the big hotel. These youngsters, he thought, hadn't the faintest idea of the peril into which he was leading them. They were as ignorant as babies of the dark evil and unearthly danger of the interplanetary frontier.

He hardened himself against the qualms of conscience. There was that at stake, he told himself fiercely, against which the safety of a lot of spoiled, rich young people was absolutely nothing.

Holk Or was chuckling as they emerged into the chill Martian night. He told Kenniston admiringly, "That was one of the smoothest jobs of lying I ever heard, that story about finding John Dark's treasure. Take it from me, it was slick!"

The Jovian guffawed loudly as he added, "What would their faces be like if they knew that John Dark and his crew are still living? That it was John Dark himself who sent us here?"

"Be quiet, you idiot!" ordered Kenniston hastily. "Do you want the whole Patrol to hear you?"

CHAPTER II
DISCOVERED

The *Sunsprite* throbbed steadily through the vast, dangerous wilderness of the asteroidal zone. To the eye, the cruiser moved in a black void starred by creeping crumbs of light. In reality those bright, crawling specks were booming asteroids or whirling meteor-swarms rushing in complicated, unchartable orbits and constantly threatening destruction.

For three days now, the cruiser had cautiously groped deeper into this most perilous region of the System. Now a bright, tiny disk of white light was shining far ahead like a beckoning beacon. It was the asteroid Vesta—their goal.

Kenniston, leaning against the glassite deck-wall, somberly eyed the distant asteroid.

"We'll reach it by tomorrow," he thought. "Then what? I suppose John Dark will hold these rich youngsters for ransom."

Kenniston knew that the pirate leader would instantly see the chance of extorting vast sums by holding this group of wealthy young people as captives.

"I wish to God I hadn't had to bring them into this," Kenniston sweated. "But what else could I do? It was the only way I could get back to Vesta with the materials."

His mind was going back over the disastrous events since the day three weeks before, when the Patrol had caught up to John Dark at last.

Dark's pirate ship, the *Falcon*, had been gunned to a helpless wreck. It had, fortunately for the pirates, drifted off into a region of perilous meteor-swarms where the Patrol cruisers dared not follow. The Patrol thought everybody on the pirate ship dead anyway, Kenniston knew.

But John Dark and most of his crew were still alive in the drifting wreck. They had fought the battle wearing space-suits, and that had saved them. They had clung grimly to the wreck as it drifted on and on until it finally fell into the feeble gravitational pull of Vesta.

Kenniston could still remember those tense hours when the wreck had fallen through the satellite swarm of meteors onto the World with a Thousand Moons. They had managed to cushion their crash. John Dark, always the most resourceful of men, had managed to jury-rig makeshift rocket-tubes that had softened the impact of their fall.

But the wrecked *Falcon* had been marooned there in the weird asteroidal jungle, with the alien, menacing Vestans already gathering around it. The ship would never fly space again until major repairs were made. And they could not be made until quantities of material and equipment were brought. Someone must go for those materials to Mars, the nearest planet.

John Dark had superintended construction of a little two-man rocket from parts of the ship. Kenniston and Holk Or were to go in it.

"You *must* be back with that list of equipment and materials within two weeks, Kenniston," Dark had emphasized. "If we stay castaway here longer than that, either the Vestans will get us or the Patrol discover us."

The pirate leader had added, "The moon-jewels I've given you will more than pay for a small cruiser, if you can buy one at Mars. If you can't buy one, get one any way you can—but get back here quickly!"

Well, Kenniston thought grimly, he had got a cruiser in the only way he could. Down in its hold were the berylloy plates and spare rocket-tubes and new cyclotrons he had had loaded aboard at Syrtis.

But he was also bringing back to Vesta with him a bunch of thrill-seeking, rich, young people who believed they were going on a romantic treasure-hunt. What would they think of him when they discovered how he had betrayed them?

That's Vesta, isn't it?" spoke a girl's eager voice behind him, interrupting his dark thoughts.

Kenniston turned quickly. It was Gloria Loring, boyish in silken space-slacks, her hands thrust into the pockets.

There was a naive eagerness in her clear, lovely face as she looked toward the distant asteroid, that made her look more like an excited small girl than like the bored, jewelled heiress of that night at Syrtis.

"Yes, that's the World with a Thousand Moons," Kenniston nodded. "We'll reach it by tomorrow. I've just been up on the bridge, telling your Captain Walls the safest route through the meteor swarms."

Her dark eyes studied him curiously. "You've been out here on the frontier a long time, haven't you?"

"Twelve years," he told her. "That's a long time in the outer planets. Most space-men don't last that long out here—wrecks, accidents or gravitation-paralysis gets them."

"Gravitation-paralysis?" she repeated. "I've heard of that as a terrible danger to space-travelers. But I don't really know what it is."

"It's the most dreaded danger of all out here," Kenniston answered. "A paralysis that hits you when you change from very weak to very strong gravities or vice versa, too often. It locks all your muscles rigid by numbing the motor-nerves."

Gloria shivered. "That sounds ghastly."

"It is," Kenniston said somberly. "I've seen scores of my friends stricken down by it, in the years I've sailed the outer System."

"I didn't know you'd been a space-sailor all that time," the heiress said wonderingly. "I thought you said you were a meteor-miner."

Kenniston woke up to the fact that he had made a bad slip. He hastily covered up. "You have to be a good bit of a space-sailor to be a meteor-miner, Miss Loring. You have to cover a lot of territory."

He was thankful that they were interrupted at that moment by some of the others who came along the deck in a lively, chattering group.

Robbie Boone was the center of the group. That chubby, clownish young man, heir to the Atomic Power Corporation millions, had garbed himself in what he fondly believed to be a typical space-man's outfit. His jacket and slacks were of black synthesilk, and he wore a big atom-pistol.

"Hiya, pal!" he grinned cherubically at Kenniston. "When does this here crate of ours jet down at Vesta?"

"If you knew how silly you looked, Robbie," said Gloria devastatingly, "trying to dress and talk like an old space-man."

"You're just jealous," Robbie defied. "I look all right, don't I, Kenniston?"

Kenniston's lips twitched. "You'd certainly create a sensation if you walked into the Spaceman's Rendezvous in Jovopolis."

Alice Krim, a featherheaded little blonde, eyed Kenniston admiringly. "You've been to an awful lot of planets, haven't you?" she sighed.

"Turn it off, Alice," said Gloria dryly. "Mr. Kenniston doesn't flirt."

Arthur Lanning, the sulky, handsome youngster who always had a drink in his hand, drawled. "Then you've tried him out, Gloria?"

The heiress' dark eyes snapped, but she was spared a reply by the appearance of Mrs. Milsom. That dumpy, fluttery woman, the nominal chaperone of the group, immediately seized upon Kenniston as usual.

"Mr. Kenniston, are you sure this asteroid we're going to is safe?" she asked him for the hundredth time. "Is there a good hotel there?"

"A good hotel there?" Kenniston was too astounded to answer, for a moment.

Into his mind had risen memory of the savage, choking green jungles of the World with a Thousand Moons; of the slithering creatures slipping through the fronds, of the rustling presence of the dreaded Vestans who could never quite be seen; of the pirate wreck around which John Dark and half a hundred of the System's most hardened outlaws waited.

"Of course there's no hotel there, Aunty," Gloria said disgustedly. "Can't you understand that this asteroid's almost unexplored?"

Holk Or had come up, and the big Jovian had heard. He broke into a booming laugh. "A hotel on Vesta! That's a good one!"

Kenniston flashed the big green pirate a warning glance. Robbie Boone was asking him, "Will there be any good hunting there?"

"Sure there will," Holk Or declared. His small eyes gleamed with secret humor. "You're going to find lots of adventure there, my lad."

When Mrs. Milsom had dragged the others away for the usual afternoon game of "dimension bridge," the Jovian looked after them, chuckling.

"This crowd of idiots hadn't ought to have ever left Earth. What a surprise they're going to get on Vesta!"

"They're not such a bad bunch, at bottom," Kenniston said halfheartedly. "Just a lot of ignorant kids looking for adventure."

"Bah, you're falling for the Loring girl," scoffed Holk Or. "You'd better keep your mind on John Dark's orders."

Kenniston made a warning gesture. "Cut it! Here comes Murdock."

Hugh Murdock came straight along the deck toward them, and his sober, clean-cut young face wore a puzzled look as he halted before them.

"Kenniston, there's something about this I can't understand," he declared.

"Yes? What's that?" returned Kenniston guardedly.

He was very much on the alert. Murdock was not a heedless, gullible youngster like the others. He was, Kenniston had learned, an already important official in the Loring Radium company.

From the chaffing the others gave Murdock, it was evident that the young business man had joined the party only because he was in love with Gloria. There was something likeable about the dogged devotion of the sober young man. His very obvious determination to protect Gloria's safety, and his intelligence, made him dangerous in Kenniston's eyes.

"I was down in the hold looking over the equipment you loaded," Hugh Murdock was saying. "You know, the stuff we're to use to dig out the wreck of Dark's ship. And I can't understand it—there's no digging machinery, but simply a lot of cyclotrons, rocket-tubes and spare plates."

Kenniston smiled to cover the alarm he felt. "Don't worry, Murdock, I loaded just the equipment we'll need. You'll see when we reach Vesta."

Murdock persisted. "But I still don't see how that stuff is going to help. It's more like ship-repair stores than anything else."

Kenniston lied hastily. "The cycs are for power-supply, and the rocket-tubes and plates are to build a heavy duty power-hoist to jack the wreck out of the mud. Holk Or and I have got that all figured out."

Murdock frowned as though still unconvinced, but dropped the subject. When he had gone off to join the others, Holk Or glared after him.

"That fellow's too smart for his own good," muttered the Jovian. "He's suspicious. Maybe I'd better see that he meets with an accident."

"No, let him alone," warned Kenniston. "If anything happened to him now, the others would want to turn back. And we're almost to Vesta now."

But worry remained as a shadow in the back of Kenniston's own mind. It still oppressed him hours later when the arbitrary ship's-time had brought the 'night.' Sitting down in the luxurious passenger-cabin over highballs with the others, he wondered where Hugh Murdock was.

The rest of Gloria's party were all here, listening with fascinated interest to Holk Or's colorful yarns of adventures on the wild asteroids. But Murdock was missing. Kenniston wondered worriedly if the fellow was looking over that equipment in the hold again.

A young Earth space-man—one of the *Sunsprite's* small crew—came into the cabin and approached Kenniston.

"Captain Walls' compliments, sir, and would you come up to the bridge? He'd like your advice about the course again."

"I'll go with you," Gloria said as Kenniston rose. "I like it up in the bridge best of any place on the ship."

As they climbed past the little telaudio transmitter-room, they saw Hugh Murdock standing in there by the operator. He smiled at Gloria.

"I've been trying to get some messages through to Earth, but it seems we're almost out of range," he said ruefully.

"Can't you ever forget business, Hugh?" the girl said exasperatedly. "You're about as adventurous as a fat radium-broker of fifty."

Kenniston, however, felt relieved that Murdock had apparently forgotten about the oddness of the equipment below. His spirits were lighter when they entered the glassite-enclosed bridge.

Captain Walls turned from where he stood beside Bray, the chief pilot. The plump, cheerful master touched his cap to Gloria Loring.

"Sorry to bother you again, Mr. Kenniston," he apologized. "But we're getting pretty near Vesta, and you know this devilish region of space better than I do. The charts are so vague they're useless."

Kenniston glanced at the instrument-panel with a practiced eye and then squinted at the void ahead. The *Sunsprite* was now throbbing steadily through a starry immensity whose hosts of glittering points of light would have made a bewildering panorama to laymen's eyes.

They seemed near none of those blazing sparks. Yet every few minutes, red lights blinked and buzzers sounded on the instrument panel. At each such warning of the meteorometers, the pilot glanced quickly at their direction-dials and then touched the rocket-throttles to change course slightly. The cruiser was threading a way through unseen but highly perilous swarms of rushing meteors and scores of thundering asteroids.

Vesta was now a bright, pale-green disk like a little moon. It was not directly ahead, but lay well to the left. The cruiser was following an indirect course that had been laid to detour it well around one of the bigger meteor-swarms that was spinning rapidly toward Mars.

"What about it, Mr. Kenniston—is it safe to turn toward Vesta now?" Captain Walls asked anxiously. "The chart doesn't show any more swarms that should be in this region now, by my calculations."

Kenniston snorted. "Charts are all made by planet-lubbers. There's a small swarm that tags after that big No. 480 mess we just detoured around. Let me have the 'scopes and I'll try to locate it."

Using the meteorscopes whose sensitive electromagnetic beams could probe far out through space, to be reflected by any matter, Kenniston searched carefully. He finally straightened from the task.

"It's all right—the tag-swarm is on the far side of No. 480," he reported. "It should be safe to blast straight toward Vesta now."

The captain's anxiety was only partly assuaged. "But when we reach the asteroid, what then? How do we get through the satellite-swarm around it?"

"I can pilot you through that," Kenniston assured him. "There's a periodic break in that swarm, due to gravitational perturbations of the spinning meteor-moons. I know how to find it."

"Then I'll wake you up early tomorrow 'morning' before we reach Vesta," vowed Captain Walls. "I've no hankering to run that swarm myself."

"We'll be there in the morning?" exclaimed Gloria with eager delight. "How long then will it take us to find the pirate wreck?"

Kenniston uncomfortably evaded the question. "I don't know—it shouldn't take long. We can land in the jungle near the wreck."

His feeling of guilt was increased by her enthusiastic excitement. If she and the others only knew what the morrow was to bring them!

He did not feel like facing the rest of them now, and lingered on the dark deck when they went back down from the bridge. Gloria remained beside him instead of going on to the cabin.

She stood, with the starlight from the transparent deck-wall falling upon her youthful face as she looked up at him.

"You *are* a moody creature, you know," she told Kenniston lightly. "Sometimes you're almost human—then you get all dark and grim again."

Kenniston grinned despite himself. Her voice came in mock surprise. "Why, it can actually smile! I can't believe my eyes."

Her clear young face was provocatively close, the faint perfume of her dark hair in his nostrils. He knew that she was deliberately flirting with him, perhaps mostly out of curiosity.

She expected him to kiss her, he knew. Damn it, he *would* kiss her! He did so, half ironically. But the ironic amusement faded out of his mind somehow at the oddly shy contact of her soft lips.

"Why, you're just a kid," he muttered. "A little kid masquerading as a bored, sophisticated young lady."

Gloria stiffened with anger. "Don't be silly! I've kissed men before. I just wanted to find out what you were really like."

"Well, what did you find out?"

Her voice softened. "I found out that you're not as grim as you look. I think you're just lonely."

The truth of that made Kenniston wince. Yes, he was lonely enough, he thought somberly. All his old space-mates, passing one by one—

"Don't you have anyone?" Gloria was asking him wonderingly.

"No family, except my kid brother Ricky," he answered heavily. "And most of my old space-partners are either dead or else worse—lying in the grip of gravitation-paralysis."

Memory of those old partners re-established Kenniston's wavering resolution. He mustn't let them down! He must go through with delivering this cruiser's cargo to John Dark, no matter what the consequences.

He thrust the girl almost roughly from him. "It's getting late. You'd better turn in like the others."

But later, in his bunk in the little cabin he shared with Holk Or, Kenniston found memory of Gloria a barrier to sleep. The shy touch of her lips refused to be forgotten. What would she think of him by tomorrow?

He slept, finally. When he awakened, it was to realization that someone had just sharply spoken his name. He knew drowsily it was 'morning' and thought at first that Captain Walls had sent someone to awaken him.

Then he stiffened as he saw who had awakened him. It was Hugh Murdock. The young businessman's sober face was grim now, and he stood in the doorway of the cabin with a heavy atom-pistol in his hand.

"Get up and dress, Kenniston," Murdock said sternly. "And wake up your fellow-pirate, too. If you make a wrong move I'll kill you both."

CHAPTER III
THROUGH THE METEOR-MOONS

Kenniston went cold with dismay. He told himself numbly that it was impossible Hugh Murdock could have discovered the truth. But the grim expression on Murdock's face and the naked hate in his eyes were explainable on no other grounds.

The young businessman's finger was tense on the trigger of the atom-pistol. Resistance would be senseless. Mechanically, Kenniston slipped from his bunk and threw on his slacks and space-jacket. Holk Or was doing the same, the big Jovian's battered green face almost ludicrous in astonishment.

"Now perhaps you'll tell us what this means," Kenniston said harshly, his mind racing. "Have you lost your senses?"

"I've just come to them, Kenniston," rapped Murdock. "What fools we all were, not to guess that you two belong to Dark's pirates!"

Kenniston's lips tightened. It was clear now that Murdock had actually discovered something. From Holk Or came an angry roar.

"Devils of Pluto, I'm no pirate!" the big Jovian lied magnificently. "Whatever gave you this crazy idea?"

Murdock's hard face did not relax. He waved the atom-pistol. "Go into the main cabin," he ordered. "Walk ahead of me."

Helplessly, Kenniston and Holk Or obeyed. His mind was desperate as he shouldered down the corridor. The throbbing of the rockets told him the *Sunsprite* was still forging through the void. They must be very near Vesta by now—and now this had to happen!

The others had been awakened by the uproar and streamed into the main cabin after Murdock and his two prisoners. Kenniston glimpsed Gloria, slim in a silken negligee, her dark eyes round with amazement.

"Hugh, have you gone crazy?" she exclaimed stupefiedly.

Murdock answered without looking toward her. "I've found out the truth, Gloria. These men belong to John Dark's crew. They were taking us into a trap."

"Holy smoke!" gasped Robbie Boone, his jaw sagging as the chubby youth stared at Kenniston and Holk Or. "They're pirates?"

"I think you must be losing your mind!" Gloria stormed at Hugh Murdock. "This is ridiculous."

Holk Or yawned elaborately. "Space-sickness hits people in queer ways, Miss Loring," the Jovian told Gloria confidentially. "Some it just makes sick, but others it makes delirious."

"I'm not delirious, and you two know it," Murdock retorted grimly. He spoke to Gloria and the others, without taking his eyes or the muzzle of his pistol off his two captives.

"I thought from the first that this Kenniston's story of finding the wreck of Dark's ship on Vesta was a thin one," Murdock declared. "And yesterday my suspicions were increased when I went down and looked over the cargo of equipment they brought. It's not equipment to dig out a buried wreck. It's equipment to *repair* a damaged ship—John Dark's ship!

"Suspecting that, last 'night' I sent a telaudiogram to Patrol headquarters at Earth. I gave full descriptions of Kenniston and this Jovian and inquired if they had criminal records. An answer came through an hour ago. This fellow Holk Or has a record of criminal piracy as long as your arm, and was definitely known to be one of John Dark's crew!"

There was an incredulous gasp from the others. Murdock still grimly watched Kenniston and the Jovian as he concluded.

"The Patrol hasn't yet sent through Kenniston's record, but it's obvious enough that he's one of Dark's men too, and that his story that he and the Jovian are meteor-miners is a flat lie."

"I can't understand this," muttered young Arthur Lanning, staring. "If they're Dark's men, why should they induce us to go to Vesta?"

"Can't you see?" said Hugh Murdock. "John Dark's ship did crash on Vesta after being wrecked—that must be true enough. But Dark and his pirates weren't dead as the Patrol thought. They had to have machines and material to repair their ship. So Dark sent these two men to Mars for the materials. The two couldn't get a ship there any other way, so they made use of our cruiser by selling us that treasure yarn!"

Kenniston winced. He knew now that he had underestimated Murdock, who had put together the evidence quickly when his suspicions were roused.

Gloria Loring, looking at Kenniston with wide dark eyes, saw the change in his expression. Into her white face came an incredulous loathing.

"Then it's true," she whispered. "You did that—you deliberately planned to lead us all into capture?"

"Aw, you're all space-struck," growled Holk Or, bluffing to the last.

Murdock spoke over his shoulder. "Call Captain Walls, Robbie."

"No need to—here he comes now!" yelped the excited youth.

Captain Walls, entering the cabin in urgent haste, had eyes only for Kenniston in the first moment.

"Ah, there you are, Mr. Kenniston!" the captain exclaimed relievedly. "I was just coming for you. We've reached Vesta! I've ordered the pilot to slow down, for I want you to pilot us through the swarm—"

The captain's voice trailed off. His eyes bulged as for the first time he perceived that Murdock was covering the two men with a gun.

"We're not going in to Vesta, captain," rapped Murdock. "John Dark and his pirates are on the asteroid—*alive!*"

Captain Walls' plump face went waxy as he heard the name of the most dreaded corsair of the System.

"Dark—living?" he stuttered. "Good God, you must be joking!"

Mrs. Milsom, her dumpy figure shivering and her teeth chattering with terror, pointed a finger at Kenniston and the Jovian.

"They're two of the pirates!" she shrilled. "They might have murdered us all in our beds! I knew this would happen when we left Earth—"

Kenniston's mind was seething with despair as he stood there with hands upraised. His whole desperate plan was ruined at this last moment.

He wouldn't *let* it be ruined! He would get this cargo of machines and materials to John Dark if it meant his life!

"Turn back at once toward Mars, captain," Gloria was saying quietly to the stunned officer. Her face was still very pale.

Kenniston, standing tense, had had an idea. A desperate chance to make a break, in the face of Murdock's atom-gun.

The captain had said that he had just ordered the pilot to slow down the *Sunsprite*. In a moment would come the shock of the braking rocket-tubes firing from the bows—

That shock came an instant after the wild expedient flashed across Kenniston's mind. It was only a jarring vibration through the fabric of the ship, for the pilot knew his business.

It staggered them all on their feet, for just a moment. But Kenniston had been waiting for that moment. As Hugh Murdock moved his gun-arm involuntarily to balance himself, Kenniston lunged forward.

"The bridge, Holk!" he yelled as he hurled himself.

Kenniston's shoulder hit the captain and sent him caroming into Murdock. The two men sprawled on the floor.

Holk Or, with instant understanding, already had the door of the cabin open. They plunged out into the corridor together.

"Our only chance is to make the bridge and grab the controls!" Kenniston cried as they raced down the corridor. "We can keep them long enough to land on Vesta—"

Hiss—*flash!* The crackling blast of the atom-gun tore into the lower steps of the ladder up which he and the Jovian frantically climbed. Murdock was running after them as he fired, and there were shouts of alarm.

Kenniston and Holk Or burst into the glassite-walled bridge. Bray, the pilot, turned for a startled moment from his rocket-throttles.

Beyond the pilot, the transparent front wall framed a square of black space in which bulked the monstrous sphere of the nearby asteroid.

The World with a Thousand Moons! It loomed up only a few hundred miles away, a big, pale-green sphere encircled by the vast globular swarm of hundreds on hundreds of gleaming little meteor-satellites.

"Why—what—" stammered the pilot, bewildered.

Kenniston's fist caught his chin, and the man sagged to the floor.

"Bar the door, Holk!" yelled Kenniston as he leaped toward the rocket-throttles.

"Hell, there's only a catch!" swore the Jovian. He braced his brawny shoulders against the metal door. "I can hold it a little while."

Kenniston's hands were flashing over the throttles. The *Sunsprite* was moving at reduced speed toward the meteor-enclosed asteroid.

The cruiser shook to the bursting roar of power, as he opened up all the tail rockets. It plunged visibly faster toward the deadly swarm around Vesta, picking up speed by the minute.

Rocking, creaking, quivering to the dangerous rate of acceleration Kenniston was maintaining, the little ship rushed ahead. But now there was loud hammering at the bridge-room door.

"Open up or we'll burn that door down!" came Captain Walls' yell.

Kenniston didn't turn. Hunched over the throttles, peering tensely ahead, he was tautly estimating speed and direction. His eyes searched frantically for the periodic break in the outer meteors.

There was a muffled crackling and the smell of scorched metal flooded the bridge-room. A hoarse exclamation of pain came from Holk Or.

"They got my arm through the door, damn them!" cursed the Jovian. "Hurry, Kenniston!"

Kenniston was driving the *Sunsprite* full speed toward the whirling cloud of meteors around the asteroid. He had spotted the break in the cloud, the periodic opening caused by the gravitational influence of another nearby asteroid.

It was not a real opening. It was merely a small area in the swarm where the rushing meteors were not so thick, and where a ship had a chance to worm through by careful piloting.

Kenniston only remotely heard the struggle that Holk Or was putting up to hold the door against the hammering crowd outside. His mind was wholly intent on the desperately ticklish piloting at hand.

He cut speed and eased the *Sunsprite* down into that thinner area of the meteor-swarm. Space around them now seemed buzzing with rushing, brilliant little moons.

The meteorometers had gone crazy, blinking and buzzing unceasing warning, their needles bobbing all over the direction-dials. Instruments were useless here—he had to work by sight alone. He eased the cruiser lower through the swarm, his fingers flashing over the throttles, using quick bursts of the rockets to veer aside from the bright, rushing meteors.

"Hurry!" yelled Holk Or hoarsely again, over the tumult. "I can't—hold them out much longer—"

Down and down went the *Sunsprite* through the maze of meteor-moons, twisting, turning, dropping ever lower toward the green asteroid.

A last gasping shout from Holk Or, and the door crashed off its burned-through hinges. Kenniston, unable to turn from the life-or-death business of threading the swarm, heard the Jovian fighting furiously.

Next moment a hand gripped Kenniston's shoulder and tore him away from the controls. It was Murdock, his eyes blazing, his gun raised.

"Raise your hands or I'll kill you, Kenniston!" he cried.

"Let me go!" yelled Kenniston, struggling to get back to the throttles. "You *fool!*"

He had just glimpsed the jagged moonlet rushing obliquely toward them from the left, bulking suddenly big and monstrous.

Crash! The shock flung them from their feet, and the *Sunsprite* gyrated crazily in space. There was a blood-chilling shriek of outrushing air from the fore part of the ship, and the slam-slam-slam of the automatic air-doors closing, down there.

The cruiser's whole bows had been crushed in by the glancing blow of the meteor. Now, ironically, the ship was falling clear of the meteor-swarm for Kenniston's piloting had almost won through it before the impact. But the *Sunsprite* was falling helplessly, turning over and over as it plunged down toward the green surface of the jungled asteroid.

My God, we're struck!" came Captain Walls' thin yell.

"This is your fault!" Murdock blazed at Kenniston. "You damned pirates will die for this!"

"Let me at those controls or we'll all die together in five minutes!" Kenniston cried. "We'll crash to smithereens unless I can make a tail-tube landing—"

Heedless of Murdock's gun, he jumped to the controls. His hands flew over the throttles, firing desperate quick bursts of the tail rocket-tubes to bring them out of the spin in which they were falling.

The brake-rockets in the bow were gone. The ship was crippled, almost impossible to handle. And the dark green jungles of Vesta's surface were rushing upward with appalling speed.

Kenniston's frantic efforts brought the *Sunsprite* out of the spin. By firing the lateral rockets, he kept it falling tail-downward.

"We're goners!" yelled someone in the stricken ship. "We're going to crash!"

Air was screaming outside the plummeting ship. Kenniston, his hands superhumanly tense on the throttles, mechanically estimated their distance from the uprushing green jungles.

He glimpsed a little black lake in the jungle, and near it the big circle of an electrified stockade. He recognized it—John Dark's camp!

Then, a thousand feet above the jungle, Kenniston's hands jerked open the throttles. The tail rockets spouted fire downward.

Sickening shock of the sudden check almost hurled him away from the controls. His hands jabbed the throttles in and out with lightning rapidity, checking their further fall with one quick burst after another.

A sound of rending branches—a staggering sidewise shock that flung him from his feet. A jarring thump, then silence. They had landed.

CHAPTER IV
THE VESTANS

Kenniston picked himself up groggily. The others in the bridge had been thrown against walls or floor by the shock, but seemed no more than bruised. Holk Or was nursing his burned arm. But Hugh Murdock, staggering in a corner, still held his atom-pistol trained on Kenniston and the Jovian.

"My God, what a landing!" exclaimed Captain Walls, his plump face still white. "I thought we were done for."

"Maybe we still are," Murdock said grimly. He said savagely to Kenniston, "You think you've won, don't you? Because you've managed to crash us on this asteroid where your pirate boss is waiting?"

"Listen, Murdock—," Kenniston began desperately.

"Keep your hands up or I'll kill you both!" blazed Murdock. "March down to the main cabin."

Kenniston and the Jovian obeyed. The *Sunsprite* was lying sharply canted on its side, and it was difficult to scramble down through the tilted passageways and decks to the big main cabin.

The cabin was a scene of confusion, for it was impossible to stand upright on its tilted floor. Young Arthur Lanning had been stunned, and Gloria Loring and the scared blonde girl, Alice Krim, were bathing his bruised forehead. Robbie Boone was peering wildly through a porthole at the sunlit tangle of green jungle outside. From Mrs. Milsom came a shrill, steady wail of terror.

"Stop that screeching," Murdock told the dumpy dowager brutally. "You're not hurt. Gloria, are you others all right?"

Gloria raised her white face from her task. "Only bruised, Hugh."

She did not look at Kenniston or the big Jovian as she spoke.

Robbie Boone's teeth were chattering. "Murdock, what are we going to do? We're wrecked, on this hellish jungle asteroid—"

Murdock paid the frightened, chubby youth no attention. Captain Walls, Bray, and four of the crew were entering the cabin. The captain and pilot had belted on atom-pistols.

Captain Walls' plump face was paler. "Two of the crew were killed and our telaudio wrecked by that meteor," he reported. He glared at Kenniston. "You damned pirate! You're responsible for this!"

"If you hadn't dragged me away from the controls, the cruiser wouldn't have been struck," Kenniston denied. "And I'm not a pirate—"

Murdock interrupted. "We'll settle with those two later," he told the enraged captain. "Right now, we'll have to get out of the ship. We can't stay in here until we get it righted on an even keel."

Holk Or rumbled a warning. "Better be careful about going outside. Those cursed Vestans are thick in these jungles."

"I'll have no advice from you two pirates!" flamed the captain. "Bray, you and Thorpe keep your guns on them every minute."

The heavy main space-door was opened. Pale sunlight and warm, steamy air laden with rank scents of strange vegetation drifted in. Outside lay a raw clearing the falling ship had crushed out of the jungle.

Captain Walls supervised as they all donned lead-soled weight-shoes to compensate for the weaker gravity. Then they emerged, young Lanning being supported by Murdock and Robbie. Kenniston and the Jovian were last to emerge, under the watchful guns of their guards.

The crew and passengers were looking around with wonder and revulsion. The silvery bulk of the *Sunsprite* lay awkwardly heeled on its side. The symmetrical torpedo shape of the cruiser was now badly marred by the crumpled condition of its bow.

All around them in the thin sunlight rose slender trees whose enormous green leaves grew directly from the trunks. This grotesque forest was made more dense by festoons of writhing "snake-vines," weird rootless creepers which crawled like plant-serpents from one tree to another. Each stir of wind brought white spore-dust down in a shower from the trees.

The few living creatures of this forbidding landscape were equally alien. Big white meteor-rats scurried on their eight legs through the brush. Phosphorescent flame-birds shot through the upper fronds like streaks of fire. In the pale sky overhead, there were ceaseless gleams and flashes of light as the spinning meteor-swarm reflected the sunlight.

"What a horrible place!" shrilled Mrs. Milsom. "We'll all die here— we'll never get back to Earth. I knew this would happen!"

"This is certainly a mean spot to be cast away," muttered Captain Walls. "God knows what queer creatures inhabit it, not to speak of the mysterious Vestans everybody talks about. And John Dark and his crew are somewhere here. And the telaudio wrecked, so we can't call for help."

Kenniston realized that none of the others had glimpsed Dark's camp as they fell. They didn't know the pirate encampment was only a few miles away in the jungle.

"What are we going to do, captain?" Gloria was asking, her face still pale but her voice quite steady. "Can we get away?"

Captain Walls looked hopeless. "We can't take off with the whole bow of the *Sunsprite* crushed in."

"We can repair it, can't we?" Hugh Murdock suggested. "Remember, in the hold is the cargo of machinery and repair-materials that Kenniston was bringing to repair Dark's ship. Can't we use that equipment?"

The captain looked more hopeful. "Maybe we can. Bray and the crew and I ought to be able to do an emergency job of patching the bow and installing new rocket-tubes there. But we'll have to work fast to get away before Dark's outfit learns we're here."

He pointed vindicatively at Kenniston. "Better lock up that fellow and his partner to make sure he doesn't signal somehow to his fellow-pirates."

Kenniston tried again to explain. "Will you all listen to me? I tell you, I'm no pirate!"

Murdock eyed him sternly. "Do you deny that John Dark sent you to Mars for repair-equipment, and that you told us that lying treasure-story to get the equipment here in our ship?"

"No, I don't deny that," Kenniston admitted. "But I'm not one of John Dark's crew—I never was! I was a prisoner on his ship, captured by the pirates before they themselves were attacked by the Patrol."

"Do you expect us to believe that?" Murdock said incredulously.

"It's true!" Kenniston insisted. "My kid brother Ricky and I were captured by John Dark's outfit several weeks ago. We were prisoners on his ship when it was wrecked by the Patrol. After the wreck drifted onto Vesta here, Dark wanted to send someone to Mars for repair-equipment. He wouldn't send one of his own men in charge, for fear the man would double-cross him and never come back.

"So he sent me, his prisoner, on that errand. Holk Or came along to help me navigate a ship back. And I had to obey Dark and get the equipment back here at any cost. For Dark kept my brother Ricky prisoner here with him, and told me that if I didn't bring back that equipment, Ricky would be shot!"

Holk Or spoke up. "It's true, what Kenniston's telling you," rumbled the Jovian. "Me, I'm one of Dark's pirates and I don't care a curse who knows it. But Kenniston did this only to save his brother."

"I don't believe it," said Captain Walls flatly. "It's another of the smooth lies this fellow Kenniston makes up so easily."

Gloria spoke to Kenniston, her dark eyes still accusing. "If what you say is true and you're not a pirate, then you brought all of us into this danger simply to save your own brother?"

Kenniston looked at her miserably. "Yes, I did. I was willing to lead you all into capture to save Ricky. But I had a reason—"

"Sure, you had a reason," Murdock said bitterly. "What did the safety of strangers like us mean to you, compared to your precious brother?"

Captain Walls motioned Kenniston and Holk Or angrily toward the ship. "Bray, take them in and lock them under guard in a cabin," he said.

Holk Or suddenly yelled. "Look out! There's a Vestan!"

Kenniston, his blood chilling with alarm, glanced where the Jovian pointed. At the west edge of the clearing, a small animal had suddenly emerged from the dense green jungle.

It was a six-legged, striped, catlike beast, not unordinary as interplanetary animals go. But its head looked queer, seeming to have a bulbous gray mass attached behind its ears.

Captain Walls uttered a scoffing exclamation. "That's only an ordinary asteroid-cat."

"That *is* a Vestan!" Kenniston cried. "Shoot at its head—"

His warning was too late. The catlike beast had launched itself in a spring toward their group.

As its striped body shot through the air, Walls triggered his atom-pistol. The crackling blast of force tore into the body of the charging asteroid-cat, and the beast fell heavily a few yards away.

But as it fell, the small gray mass upon its neck suddenly detached itself from the dead animal and scuttled swiftly forward. It moved with blurring speed toward Bray, the nearest to it of the group.

The little gray creature was no bigger than a man's clenched fists together. It was a gray, wrinkled featureless thing, except for pinpoint eyes and the tiny clawlike legs upon which it scurried. It reached Bray and ran swiftly up his legs and back as he swore startledly.

Kenniston, made reckless of danger by his horror, yelled and lunged toward the pilot. Bray was swearing and trying to slap at the gray thing running up his back. But the little creature had now reached his neck. Clinging there, it swiftly dug two tiny, needle-like antennae into the base of his neck.

"Hold him!" Kenniston shouted hoarsely. "The Vestan has got him!"

Bray had undergone a sudden metamorphosis as the gray creature dug its antennae into his neck. His face stiffened, became masklike.

The pilot turned and began to run stiffly toward the jungle. Kenniston's leap almost caught him, but Bray lashed out a fist that sent Kenniston sprawling.

"Don't let him get away!" Kenniston yelled, scrambling up.

But the others were too stricken by amazement and horror to interfere in time. Bray had already plunged into the jungle and was gone.

"My God, what happened?" Captain Walls exclaimed dazedly. "Bray went clean crazy!"

His gun was pointing at Kenniston and Holk Or as though he held them responsible for what had occurred.

"He didn't go crazy, but he's lost now," Kenniston said heavily. "That little gray creature was one of the Vestans."

"But what did it *do* to him? That thing wasn't big enough to harm anybody."

"That's all you know about it," said Holk Or ominously. "Those little Vestans are the most dangerous creatures in the System."

"The Vestans," Kenniston added dully, "are semi-intelligent *parasites*. The live by attaching themselves to and taking control of some other creature's body. They do it by jabbing in those tiny, needle-like antennae to contact the victim's nervous system. Thereafter, the Vestan controls the victim's body absolutely. When the victim dies or is hurt, the Vestan simply detaches himself and fastens upon a new victim."

Horror was on the white faces of the others. Murdock gulped and asked, "Then Bray—"

"Bray is beyond saving now," Kenniston said. "The Vestan parasite will control his body till he dies. The Vestans always like to attach themselves to human beings—they know that a man's body is more versatile in its capabilities than an animal's."

Twilight was beginning to descend upon the little clearing in the jungle, for the sun had gone down during the last few minutes. In the gathering dusk, the jungle loomed dark and brooding about them.

Overhead, the sky of this World with a Thousand Moons was burgeoning into its full glory. The hundreds of meteor-moons that spun across the heavens were shining brighter and brighter in the deepening dusk.

Captain Walls broke the spell of horror and dread. "We'd better get back inside the ship for tonight," he said nervously. "We can't do anything about repairs until tomorrow, anyway. By then we'll have figured out some way to deal with those devilish creatures."

Murdock said bitterly to Kenniston, "Bray's end is your fault, Kenniston. You brought him and us and these women into this place, all for the sake of that brother of yours."

"He'll stand trial for that when we get back to Mars," the captain vowed. "Even if he wasn't one of Dark's crew originally, by helping them he's made himself a space-pirate, liable to execution."

Kenniston made no attempt to defend himself. He knew they wouldn't understand why he had sacrificed them for Ricky's sake, even if he told them.

He and Holk Or were locked in one of the little cabins, after it had been carefully searched. The crewman Thorpe was stationed as a guard outside their bolted door.

Holk Or, who had bandaged his burned arm, looked around the dark little cabin disgustedly. "This is a devil of a fix to get into!" swore the Jovian. "Here we've reached Vesta with the stuff, but can't let the chief know."

Kenniston asked him earnestly, "Holk, would John Dark really shoot Ricky if I didn't deliver the equipment? He said he would, but you know he needs Ricky."

Kenniston was clinging to this last shred of hope for his brother. John Dark and his pirates did need Ricky. For Ricky was a physician—Doctor Richard Kenniston of the Institute of Planetary Medicine.

That was why John Dark had spared the lives of the two brothers when he had captured them in the freighter in which they were returning to Earth from Saturn. Ordinarily, the pirate leader would have ruthlessly killed them as he killed all prisoners who were not rich enough to pay ransom.

But the fact that Ricky was a physician had saved them. The pirates needed a doctor. They had kept the two brothers prisoner on their ship for that reason. Kenniston and Ricky had still been on the *Falcon* as prisoners, when the Patrol had finally caught up to it and wrecked it.

"Dark knows that Ricky is a fine doctor and he needs a doctor," Kenniston repeated hopefully, to the Jovian. "Surely he wouldn't be foolish enough to shoot Ricky, even if I don't deliver the equipment."

"Kenniston, don't fool yourself," warned Holk Or. "The chief said he'd shoot him if you weren't back with the stuff in two weeks, and shoot him he will. John Dark never breaks his word."

That assurance sank the iron deeper into Kenniston's tormented soul. If that was true, and he knew in his heart it was, Ricky would die two days from now unless he'd delivered the repair-equipment to Dark.

He mustn't *let* Ricky die! Too much depended on his young brother's life. He must save Ricky even if it did mean the capture of Gloria and the others by the pirates. Better that they be held for ransom, than for Ricky to be killed!

Kenniston got to his feet, rigid with decision. "Then we've got to get out of here," he muttered. "We've got to escape and take word to Dark that the equipment is here."

He continued quickly, "Holk, Dark's camp is only a few miles north of here. I spotted it as the *Sunsprite* fell."

Holk Or uttered an exclamation. "Why the devil didn't you tell me so! I figured it was on the other side of the asteroid, maybe, and that we'd never find it in the jungle even if we did get away."

"It still won't be easy for us," Kenniston warned. "The Vestans may get us in the jungle between here and Dark's camp. And anyway, how can we get out of this cabin?"

The big Jovian grinned. "That'll be easy. I'd have been out of here before now, only I was waiting for the ship to quiet down."

Kenniston stared. "That door is bolted. And there's no tool or weapon in the cabin. They didn't forget a thing when they searched it!"

Holk Or's grin deepened. "They forgot one thing. They forgot how strong a Jovian is on a little, weak-gravity asteroid like this!"

CHAPTER V
NIGHT ATTACK

Kenniston caught desperately at the hope implied by the Jovian's words.

"What do you mean, Holk?"

"I mean that I'm a hundred times stronger on this little asteroid than I am on my own world, Jupiter. I can break the bolt of that door any time I want to."

"But there's an armed guard stationed outside it."

"I know, and that's where you come in, Kenniston. When I rip the door open, you be ready to jump the guard."

Kenniston considered swiftly. The chance of their getting out of the ship and safely through the jungles to the pirate camp, even if they escaped this cabin, seemed a slim one. Yet it presented the only possibility of delivering the equipment in the hold to John Dark.

The bitter irony of it struck Kenniston, for the hundredth time. He, Lance Kenniston, honorable space-man for a dozen years, working desperately to aid the most notorious pirate in the void! Even drawing into danger the girl for whom he felt—

He shut Gloria out of his mind. He mustn't think of her now. He must think only of Ricky, and of what would be lost if Ricky died. He must risk everything, sacrifice everything, to prevent that loss.

"We might as well try it now," he told the Jovian in low tones. "The ship seems quiet."

"I'll do my best to make as little noise as possible," Holk Or muttered. "Are you ready?"

The Jovian's big hands grasped the knob of the door. Kenniston crouched a little behind him, every muscle tense.

Holk Or suddenly put all his gigantically magnified strength into a tremendous tug at the door. Its bolt snapped with a crack like that of a pistolshot, and it swung wide open.

The man on guard outside turned startledly, his hand darting to the atom-gun at his belt and his mouth open to yell. But Kenniston had launched himself like a human projectile as the door was torn open.

Kenniston's fist smashed the space-sailor's chin and the man sagged limp and unconscious with no chance to utter the cry on his lips. Hastily, Kenniston took his atom-pistol and eased him to the floor.

He and Holk Or listened tensely. The single sharp crack of the snapping bolt had apparently aroused no one. The ship was silent. All aboard were sleeping exhaustedly.

"Come on," Kenniston murmured tensely to the Jovian. "We've got to hurry to get to Dark's camp before night is over."

Holk Or chuckled. "The chief will welcome us with open arms when he learns we've got the equipment here for him."

Kenniston gripped the atom-pistol as they stole through the dark ship and out of the space-door. Outside, they paused in the darkness.

The scene was one of magic, unearthly beauty. The metal bulk of the cruiser and the towering jungle around the clearing were washed by brilliant silver light that fell from the wonderful night sky of this World with a Thousand Moons.

A thousand moons indeed seemed blazing in the canopied heavens overhead! The whole dark sky was crowded by the shining moonlets that rushed ceaselessly across the firmament with the spinning of the meteor-swarm of which they were part. It was like the glorious vista of a world seen in dreams.

But Kenniston was familiar with the unearthly spectacle. He led the way rapidly toward the northern edge of the jungle.

"We'll just have to plunge in and head north," he told the Jovian. "If we reach that little lake, we can soon find Dark's camp."

They started into the dense jungle, a fairyland of silver beams sifting through the choking fronds. Something scurried close by.

"Kenniston, shoot!" cried Holk Or instantly.

Kenniston had already glimpsed the white beast scurrying toward them across a little patch of moonlight. It was one of the big meteor-rats. On its neck bunched one of the little gray masses—a Vestan.

The horror inspired by the hideous parasites tightened Kenniston's finger convulsively on the trigger of the atom-pistol. The crackling bolt of fire from the weapon ripped into the Vestan on the meteor-rat, and both parasite and animal victim were instantly a scorched, smoking heap.

"Hell, that's torn it!" cried the big Jovian. "We've roused the whole ship!"

Men awakened by the blast of the atom-gun were pouring out of the *Sunsprite,* rushing after the two escaped men. Kenniston heard Captain Walls shouting.

"They're in the jungle here! Spread out and surround them!" the officer was ordering.

Kenniston and the Jovian plunged forward, seeking to escape northward. But they had come up against an impenetrable abatis of brush.

Before they could find a way around it, they heard men crashing all around them. They were completely encircled.

"Kenniston, you and that Jovian walk back into the clearing with your hands raised or we'll blast every inch of the brush till we get you!" came the stentorian shout of the captain.

"The devil—they've got us boxed!" exclaimed Holk Or furiously. "We'll try to fight our way through."

"No!" Kenniston declared. "We couldn't make it anyway. And I'm not going to shoot innocent men."

Holk Or angrily grabbed for the atom-pistol, but Kenniston promptly threw it away. Not even in this last extremity could he bring himself to kill.

"You're a fool!" gritted the Jovian. "Now there's nothing for it but surrender."

With their hands raised, they walked out of the jungle into the brilliant silvery light of the clearing. Instantly they were surrounded by Captain Walls, Murdock and the other armed crew-men.

The girls and their scared chaperon, and young Lanning and Robbie Boone, were emerging in alarm from the *Sunsprite.* Kenniston did not look toward them.

Captain Walls' face was grim in the moonslight, as he and his men covered the two captured fugitives. "Kenniston, you and this Jovian were going to make your way to John Dark and tell him of our presence here, weren't you? You needn't deny it—it's plain enough."

"Sure we were!" exclaimed the angry Jovian. "We'd have made it, too, if a Vestan hadn't jumped us in the jungle."

"That would have meant capture of us all by Dark's pirates," said the captain grimly. "You two are a danger to us all, while you live. I'm going to remove that danger. As master of a space-ship, I have legal right to order summary execution of any space-pirates I capture. I'm going to order that now."

"You're going to kill them?" exclaimed Gloria. "Oh, no—you can't!"

"It's absolutely necessary, before they betray us to the pirates, Miss Loring," defended the captain. "They'd be sentenced to death by the courts if we took them back to Mars, anyway. But we daren't take a chance on keeping them prisoned that long."

"But just to shoot them down!" said Gloria horrifiedly. "I won't stand for that!"

Murdock took her by the arm. "It's space law, Gloria," he told her earnestly. "You'd better go back into the ship."

Kenniston stood silent in the moonslight, for he realized from the finality of Walls' voice that appeals would be utterly useless. There was no use trying again to explain why he'd been willing to betray them all to save Ricky. Even if they listened, they wouldn't understand.

He felt tired, crushed, old. He'd gone a long way in the last dozen years, but every mile of it had only led toward this ending. He was going to die here under the hurtling meteor-moons of Vesta, and that meant that Ricky and Ricky's dream were going to die soon too.

"I *told* you you were a fool to throw away that gun," Holk Or was muttering.

You two march over there to the edge of the clearing," Captain Walls ordered grimly, gesturing with his gun. "Anything you want to say first, Kenniston?"

"Nothing that you would listen to or understand, you people," Kenniston answered dully. "No, I've got nothing to say."

A crackling voice came out of the dark jungle at that moment.

"*I* have something to say! Drop those guns, every man of you, and get your hands up!"

Walls spun around with an oath, levelling his atom-pistol. But out of the jungle crashed a streak of fire that hit the captain's arm and sent him reeling.

One of the girls screamed. Another of the *Sunsprite's* crew had tried to aim his weapon and had been cut down by a second bolt of atomic fire that had hit his leg.

"I *don't* want to kill you unless you force me to," came that crisp voice from the darkness. "You have ten seconds to drop the guns."

"That's the chief, Kenniston!" yelled Holk Or excitedly. "It's John Dark himself!"

The dreaded name of the pirate, a synonym for cold ruthlessness, reinforced the threat from the darkness.

Murdock let his weapon fall and shouted, "Drop the atom-guns, men! If we try to fight, the women will be hurt!"

The *Sunsprite's* men dropped their atom-pistols. Instantly out into the brilliant light from the jungle rushed a score of armed pirates. Martians, Earthmen, Venusians and others—this horde represented the criminal under-world of every planet in the System.

In a moment they had those in the clearing completely disarmed and lined up against the ship. All except Holk Or, who was loudly greeting his pirate comrades.

Kenniston saw John Dark coming across the moonslit clearing toward them. The notorious pirate was a tall, bulky Earthman, but he walked with the lightfootedness of a cat in his moonshoes. His black hair was bare, and in the silver light his black-browed, intelligent face was coldly calm as his eyes searched the row of prisoners.

"So you finally got here, Kenniston. What about the repair-equipment?" he asked sharply.

Kenniston nodded toward the *Sunsprite*. "It's in the hold. We got everything you listed."

"Good!" Dark approved. "We saw your ship crash-landing today, and started this way at once. We've been beating through the jungle, fighting off the damned Vestans, until we heard the uproar going on here. What happened? Who are these people?"

Kenniston explained briefly how he had induced Gloria Loring's party to come on a pretended treasure-hunt. He was careful to stress the wealth of the party, and John Dark reacted as he had expected.

"If they're that wealthy, their families can pay big ransoms. You've done very well, Kenniston."

"What about Ricky?" asked Kenniston tensely. "He's all right?"

"Sure he's all right—he's up at the camp," Dark answered.

Gloria said bitterly to Kenniston, "You can congratulate yourself. You've managed to save your brother."

John Dark addressed her. "Miss Loring, I presume you and your companions are willing to pay ransom for your crew also? I never take prisoners, unless they promise a good profit."

"Yes, of course we'll pay the ransom of the crew!" Gloria agreed hastily.

"Good!" said the pirate calmly. "You'll not find your captivity any more irksome than necessary."

Mrs. Milsom, the dumpy chaperon, was goggling at the notorious pirate in an extreme of terror. A sardonic gleam came into Dark's eyes as he glanced at her.

"You're a handsome wench," he told the plump dowager with mock admiration. "I've half a mind to keep you and let the ransom go."

"No, no!" shrieked the terrified woman.

Dark burst into a roar of laughter. "All right, my shrinking beauty, we'll accept ransom for you."

He turned and shot efficient orders to his subordinates, who by now had gathered behind him.

"Get that stuff out of the hold, rig up power-sledges, and start freighting it up to the camp. You'll have to cut a path through the jungle—use atom-blasters to burn one out."

One of the pirates, a hard-faced Martian, said uneasily, "That will make a racket that'll bring every Vestan on the asteroid down on us."

"You can keep the Vestans off if you keep your eyes open," Dark retorted. "Get to work, now! We've got to get the stuff up there and repair the *Falcon* at once. I'll take these prisoners up to camp."

Kenniston was grouped with the other prisoners. With a strong escort of armed pirates guarding them, and Dark and Holk Or ahead, they started through the jungle toward the pirate camp.

CHAPTER VI
ASTEROID HORROR

The pirate encampment was a big clearing hacked from the jungle a mile west of the little lake. In this space lay the long, looming black mass of the most dreaded corsair ship ever to sail the void. The *Falcon* had been righted to even keel, but its crippled condition was evident in the fused, wrecked condition of its tail rocket-tubes.

The whole camp was enclosed and protected by a shimmering blue dome of electric force. This emanated from a heavy copper cable that completely encircled the clearing, and which drew its power from insulated cables that led into the ship to generators driven by the few cyclotrons still functioning. This protective electric wall had been set up at John Dark's orders to keep out the dreaded Vestans.

John Dark raised his voice as he and his men with their prisoners approached the shimmering wall of the camp.

"Kin Ibo! Drop the wall for us!"

They saw the hard-looking Martian who was Dark's second-in-command dive into the ship to turn off the power of the electric barrier. It died, and Dark's party entered the clearing. Then the electric wall sprang into being again behind them.

Kenniston looked swiftly around. There were a score more of the motley pirates here in the camp. Also, near the side of the looming black *Falcon*, were the small, rough log huts that Dark's men had constructed.

Dark's black eyes were triumphant as he told his Martian lieutenant, "Kenniston and Holk Or brought back the equipment all right, and also brought some people who'll bring big ransom. Their wrecked ship is a few miles south. You go down there with half the men here and help the others bring up the equipment."

Kin Ibo, looking a little apprehensively out at the jungle, obeyed. Dark motioned Kenniston and the other captives toward one of the huts by the big ship.

"That hut will be your quarters until we get the *Falcon* repaired," declared the pirate leader. "Any of you who try to leave it will be shot at sight. I hope you'll not be foolish enough to attempt escape."

"That's right, folks, you wouldn't have a chance," Holk Or told them earnestly. "Even if you could get out through the electric wall, the Vestans would get you. They're thick in the jungle around here."

They silently entered the hut. Its broad open windows admitted enough of the dazzling moonslight to brighten its interior.

A dark, eager-looking young Earthman sprang up as they entered, and rushed to pump Kenniston's hand.

"Lance, you got back safely!" he exclaimed. "Thank the Lord—I've been worrying myself almost crazy about you."

"How about you, Ricky?" Kenniston asked his young brother anxiously. "You're all right?"

Ricky Kenniston nodded quickly. "Sure, I'm okay. But things haven't been so good here, Lance. The Vestans have got a half-dozen pirates who ventured outside the wall in the last few days. These creatures literally haunt the jungles around here now—I think they've been drawn here from all over the asteroid."

Ricky looked wonderingly at Gloria and the others who were entering the hut. "Lance, who are all these people? Are they prisoners of Dark too?"

"Yes, we're prisoners," Hugh Murdock told him bitterly, with a savage glance at Kenniston. "We're prisoners because your brother sacrificed us all to get back here and save *your* neck."

"Lance, you didn't do that?" Ricky exclaimed in distress.

"I had to, Ricky," Kenniston protested. "It meant your life if I didn't."

"Of course," Murdock agreed ironically. "What importance are we, compared to saving your young brother's life?"

Kenniston spoke slowly, to Murdock and Gloria and the others. "It wasn't merely Ricky's life at stake that made me sacrifice you all. It was more than that. I tried to tell you before, but you wouldn't listen."

Kenniston went across the hut and brought back the square black medicine-case of his young physician-brother. He opened it, and out of the vials and instruments inside he took a square bottle of milky fluid.

"This is what I sacrificed everything to save," Kenniston said simply.

They all stared. "What is it?" Gloria asked, puzzled.

"It's Ricky's discovery," Kenniston said. "It's a preventative and cure for gravitation-paralysis."

Captain Walls, himself an old-time space-man, was first of the group to appreciate the significance of the statement. The captain gasped.

"A preventative for gravitation-paralysis? Kenniston, are you *sure?*"

Kenniston nodded gravely. "Yes. Ricky had been working on the problem a long time, back in the Institute of Planetary Medicine. He thought he'd found a way to prevent gravitation-paralysis, the most awful scourge of all the outer System, the thing that's doomed so many space-men. But his formula required rare elements found only in the outer planets.

"Ricky and I," he continued, "went out there and secured those elements. He made up this formula, and tried it on a gravitation-paralysis case—a space-man who's lain paralyzed for years. The formula was designed to strengthen the human nervous system against the shock of varying gravitations, to re-establish an already damaged nerve-web. And it worked."

Kenniston's voice was husky as he concluded. "It worked, and that living log became a man again. The formula was a success. Ricky and I started back for Earth, where he intended to announce the discovery and arrange for its manufacture on a big scale. But, on the way back, Dark's pirates captured us."

Kenniston flung out his hand in a tortured gesture. "*That's* why I went to any lengths to save Ricky's life! It's because Ricky is the only person who knows the intricate formula of this serum. If he were to die, the secret of the cure would die with him. And that would mean that thousands on thousands more of space-men would be stricken into living death by gravitation-paralysis in the future, just as so many thousands of old friends and shipmates of mine have been stricken in the past!"

Captain Walls was the first to speak. Quietly, the plump master of the *Sunsprite* extended his hand.

"Kenniston, will you shake hands with me? And will you forgive me for everything? You did absolutely right. I'm an old space-man and I *know* what gravitation-paralysis is."

Gloria's dark eyes were glimmering with tears. "If we'd only known," she murmured to Kenniston. "No one could blame you for sacrificing a lot of worthless idlers like us, for a thing like this."

"But you're going to be all right—all of you," Kenniston assured her. "John Dark will make you pay a big ransom, but you can afford that and you'll get back safely to Earth."

"Thank Heaven for that!" exclaimed Mrs. Milsom. "I can't understand all this scientific talk of yours, but I do know that that pirate chief means no good to me. Didn't you see the lustful looks he gave me?"

The laugh that greeted this lessened the tension. Kenniston turned as Ricky plucked at his arm.

"What about ourselves, Lance?" Ricky asked quietly. "Dark still won't let us go, you know. He still needs me as a doctor."

Hugh Murdock stepped forward. "Dark would let you both go, for a big enough ransom. I'd like to pay it for you."

The handsomeness of Murdock's gesture moved Kenniston. He was only able to mutter his thanks.

While Ricky was treating Captain Walls' burned arm, the officer kept looking fascinatedly at that square bottle of milky fluid.

He said hesitantly, "I've a son—back on Earth. For five years he's lain in a cot from the gravitation-paralysis that hit him out on Jupiter. Do you suppose—"

Ricky nodded. "Yes, Captain. I'm sure that we can cure him, now."

There was an uproar out in the clearing. Kenniston went to the door and looked out.

The electric wall had temporarily been dropped, and Kin Ibo and the main body of the pirates were hastily entering the camp with their improvised power-sledges that bore heavy loads of machinery and materials.

Kenniston heard Kin Ibo reporting shrilly to John Dark, "We lost two men to the Vestans on the way here—and nearly lost two more! All this activity has drawn them from all over the asteroid! Look at that!"

Outside the electric wall, which had been hastily re-raised, could be glimpsed the shapes of lurking asteroidal animals. Meteor-rats, big striped cats, flame-birds—and every one of those lurking animals bore attached to its neck one of the little gray Vestan parasites.

John Dark was saying harshly, "We've got to have the rest of those materials to repair the *Falcon*."

"I tell you, it'd be suicide to try another trip through those jungles!" expostulated the Martian. "Those Vestans are devils!"

"Bah, you Martians are all alike—no good when your superstitions get aroused," snorted Dark contemptuously. "I'll take the men down myself. Come on, men—unload those sledges and we'll go back to the wreck."

His indomitable personality drove the scared, unwilling pirates into the task. Again the electric wall was faded out for a moment to let them out.

When they returned some time toward morning, Kenniston heard the crash of atom-guns heralding their approach. And when the wall was momentarily dropped, John Dark and his men stumbled into the camp with their loaded sledges in sweating haste.

"Turn on the wall again—quick!" bellowed Dark's bull voice. "The jungle's swarming with the gray devils now—they got five of us on the way back!"

Ricky, looking over Kenniston's shoulder, spoke appalledly. "Good God, Lance—look at them! I didn't know there *were* so many Vestans!"

Outside the barrier of shimmering electricity, scores of animals and birds dominated by the dreaded little gray parasitical creatures were now swarming. And their number seemed growing every minute.

"All this activity of the night has drawn the Vestans from far and wide," Kenniston muttered. "I don't like it. If that electric wall should fail, the creatures would be in on us in a moment."

Dark himself seemed to feel something of the same apprehension, for he was shouting urgent orders. "Hook up those atomic welders, and start putting the new plates into the *Falcon's* tail. Kin Ibo, have your gang fit in the new rocket-tubes. I'll see to installing the new cycs. If we work, we can get the job done by tomorrow night and get out of here."

Through the day, the pirates toiled with an energy that showed their earnest desire to leave the asteroid. That desire was reinforced by the ever-larger number of Vestans that now swarmed outside the wall.

There were literally hundreds of the gray parasites now outside the barrier. To have tried going outside the wall now would have been sheer suicide. The creatures were apparently driven by unholy eagerness to possess themselves of human bodies.

Gloria, looking out with Kenniston, shuddered deeply. "This horrible world! It's like a nightmare."

"We'll soon be away from it," Kenniston reassured. "See, they've almost finished repairing the *Falcon*."

The urgent toil of the pirates was showing results. By the time night came again, and the meteor-moonlets blazed forth with magic beauty in the dark heavens, the task of repair was almost done.

Kenniston and his companions had not ventured forth from the hut. Pirates were everywhere in the clearing, and all had heard John Dark's strict order to blast down the captives if they left their prison.

But from the hut, Kenniston and the others could see that the horde of Vestan-dominated animals around the camp had further increased. With ghastly avidity, they kept circling the shimmering, electric wall.

Kenniston turned in alarm at a ripping sound from the back of the log hut. Two of the logs were being torn out bodily. The battered green face and giant shoulders of Holk Or came through the opening.

"Kenniston, I came in this way because I didn't dare let Dark see me talking to you!" the Jovian exclaimed. His face was urgent in expression. "I've found out that Dark doesn't mean to let your friends here get away from Vesta alive."

"What?" exclaimed Kenniston. "That's impossible! Dark said he was going to hold Gloria and the others for ransom."

Holk Or nodded hastily. "I know, and he meant it, then. But since then, he's found out something that's changed his plans. He found it out from me—like a big fool, I told him everything when he questioned me."

The Jovian continued rapidly. "I told him that Murdock had sent that telaudio message back to Patrol headquarters, asking about my record. Now Dark figures that the Patrol will come out here to find out if that message meant that some of John Dark's outfit had actually escaped.

"Dark wants the Patrol to keep thinking that he and his outfit were destroyed—so he can slip out to Pluto and prepare a new base. So Dark, when he leaves here, is going to drop Miss Loring and her friends by the wrecked *Sunsprite*, so the Patrol will find 'em dead by the wreck and will believe their cruiser crashed accidentally. That way, they won't go on searching as they would if Miss Loring's party was all missing. And Dark will have a chance to get out to Pluto without an alarm going out."

Kenniston was suspicious. "Why do you tell us this, Holk? You're one of the pirates yourself."

"I know, but I'm afraid Dark means to drop *me* with the others by the *Sunsprite*!" Holk Or exclaimed. "He didn't say so, but I believe he figures on doing it so that the telaudio inquiry about me would be explained when I was found dead with the others by the wreck."

Murdock said swiftly, "The Jovian's right, Kenniston. All this is just what Dark *would* do, to hide his trail, now that he knows my telaudio message may have aroused the Patrol's suspicion."

Holk Or said emphatically, "I'm with you if you can figure out any way to take the *Falcon*, Kenniston!"

Kenniston paced to and fro. His whole mind was suddenly in a wild turmoil of stark fears. This meant death for Gloria and the others, and the ultimate responsibility for that death would be his.

"There is one possible chance for us to take the *Falcon*," he muttered finally. "But my God, it seems like an insane idea—"

"Wait a minute!" Captain Walls interrupted. "Dark won't drop you and your brother to die, Kenniston. He still needs your brother as a physician. You two will be safe even if we are killed."

"What of that? I can't let Gloria and the rest of you be murdered! I was willing to sacrifice you when I thought it was only a question of your being held for ransom, but this changes everything," Kenniston said wildly.

"It doesn't change anything," the captain said firmly. "Your duty is to keep your brother alive at all costs, to save that formula that means life and hope for thousands of gravitation-paralysis victims like my son."

"You mean—I should let you all be killed so Ricky and I can be saved?" Kenniston cried. "I'm damned if I will!"

"We'll never do that!" Ricky Kenniston agreed warmly. "No formula in the world is worth that."

"*This* formula is," Gloria said earnestly to Kenniston. "The captain is right."

"I won't do it," Kenniston repeated. "I have an idea by which we might be able to take the *Falcon*. We're going to try it."

"Be reasonable, Kenniston," pleaded Hugh Murdock. "None of us except Holk Or has a weapon. What chance would we have against half a hundred armed pirates?"

Kenniston looked at his brother. "Ricky, your formula strengthens the nervous system against any form of shock or damage, doesn't it? You said it did it by sheathing the nerves themselves with an impenetrable coating."

Ricky nodded puzzledly. "Yes, that's the principle. But how is that going to help us?"

"The Vestans," Kenniston reminded, "seize control of their victims by inserting those tiny needle antennae of theirs into the victim's nerve-system to establish contact. Wouldn't your formula insulate the nerves against such contact? Wouldn't it make a man immune to Vestan attack?"

"Why, it would!" Ricky declared wonderingly. "I never thought of it, yet it's entirely logical."

"Then," Kenniston said swiftly, "I want you to give every one of us, including yourself, an injection of the formula right now."

The driving purpose in his voice brushed aside all their bewildered questions and objections. Hastily, Ricky prepared his hypodermics and rapidly made an injection of the milky fluid into the big nerve-centers in the neck of each of them. Kenniston did the same for Ricky himself.

"We *should* be immune now to Vestan attack," Kenniston said prayerfully.

"But what good's that going to do us?" Holk Or demanded. "Are you figuring to try an escape into the jungle?"

"No, I'm figuring on taking the *Falcon*—by using the Vestans," Kenniston replied. "Holk, can you get into the ship and turn off the power that keeps the electric wall going? Can you drop the wall?"

The Jovian's jaw dropped. "Why, sure, I could do that, but if I did, all those hordes of Vestans outside the wall will burst in here—"

He stopped, his eyes bulging. "Good God, then that's your plan? To let the Vestans in?"

"That's it," Kenniston said tightly, his face grim. "To let the Vestans in on the pirates. That'll give us a chance to take the ship—if the formula really makes us immune to the Vestans."

The terrible nature of the proposal stunned them all. But in a moment a flame of purpose lit in the Jovian's eyes.

"I'll do it!" he swore. "It's better than waiting for Dark to kill me like he's planning. You be ready!"

The Jovian slipped out of the opening in the back of the hut. They saw him presently, casually approaching the door of the *Falcon*.

John Dark stood, a tall, dominant figure in the moonslight, barking orders to the scores of pirates who were bolting in the last of the new rocket-tubes. Kenniston's eyes swung toward the shimmering electric wall, and the horde of Vestan-dominated animals outside it.

The wall suddenly died! And as the electric barrier vanished, into the clearing came rushing the swarm of asteroidal animals.

"The wall's down!" John Dark yelled, his atom-gun leaping into his hand. "Get back into the ship — get back — "

The crash of his atom-gun drowned his own shout. Other pirates were firing wildly at the hideous creatures assailing them.

For the little gray Vestans had detached themselves from their animal victims and were swarming upon the pirates, clambering with blurring speed up their legs and backs, sinking into their necks the tiny antennae.

Kenniston glimpsed John Dark, with a hideous little gray bunch now fastened to the back of his neck, drop his gun and stalk stiffly away toward the jungle. His face was an unhuman, lifeless mask — he was a human automaton, dominated utterly by the alien creature.

"Come on!" Kenniston yelled to his friends. "Now's our chance to get into the ship!"

They plunged out of the hut into the gruesome melee. Screaming pirates were now running into the jungle in vain effort to escape the hordes of Vestans. More than half the corsairs were now overcome.

Kenniston heard a scream from Gloria as they ran, felt a swift scurrying up his back, then the needle-like stab of antennae sinking into his neck.

But the parasitic creature did *not* overpower his will! He reached around, grasped and tore loose the hideous little thing, and with strong revulsion flung it to the ground.

"Your formula works, Ricky — we're immune to them!" he gasped. "But hurry!"

Other Vestans were clambering up on them like ghastly gray spiders as they ran, but were powerless to overcome them. They tore away the creatures and plunged on.

Holk Or appeared in the door of the *Falcon*, his green face blazing as his atom-pistol pumped crashing fire into pirates inside the ship.

"I've got the ship cleared of them!" the Jovian shouted to Kenniston. "Let's get out of here!"

It was time they did so. Almost the last of John Dark's pirates had been possessed by Vestans and had become parasite-dominated robots stumbling off into the jungle. The remaining swarms of gray creatures were scurrying toward Kenniston's group.

They tumbled into the *Falcon* and slammed shut the space-door. The ship, completely if roughly repaired, was ready for take-off. Captain Walls and the men of the *Sunsprite* crew hastily started the newly-installed cyclotrons while Kenniston and the others raced up to the bridge.

Kenniston took the controls. He sent the big black pirate ship leaping up into the darkness upon flaming keel and tail-jets, and then it climbed steeply toward the wonderful sky of countless rushing moonlets.

By the time an hour had passed, the *Falcon* had groped out through the periodic break in the meteor-swarm around the asteroid. And it was throbbing at steadily increasing speed out into the vault of space, away from the World with a Thousand Moons.

"We'll head for Mars," Kenniston told the others. "We can report there to the Patrol."

"If you don't mind," Holk Or put in hastily, "I'd just as soon you dropped me at some asteroid before then. I've no desire to meet the Patrol."

Captain Walls told the Jovian, "Nonsense! After what you've done, you'll get a full pardon from the Patrol."

"You can count on it," Hugh Murdock told the doubtful Jovian. "We have some influence, back at Earth."

"Well, I guess I'll have to go honest, then," sighed Holk Or. "All the real pirate outfits are gone now, anyway." He shook his head heavily as he walked away. "The System sure isn't what it used to be."

Captain Walls was asking Ricky earnestly, "You're quite sure your formula will cure my son? All these years, I've hoped and prayed—"

"I'm certain," Ricky smiled. "Within a few weeks after we get back to Earth, gravitation-paralysis will be a thing of the past."

They moved off with the others. But Gloria lingered in the bridge with Kenniston.

"Where will you be going, after we get back?" she asked him quietly.

"Oh, back to space," he answered, a little uncomfortably. "There's nothing to hold me on Earth now that Ricky's work has succeeded."

"Nothing to hold you on Earth?" Gloria repeated. "That, I would say, is about the most ungallant speech on record."

He flushed. "You don't mean—that night on the *Sunsprite*—you weren't in earnest, surely—"

"Your passionate proposal is accepted," Gloria said calmly.

Kenniston was aghast. "But I didn't propose! I mean—I do love you, and you know it, but you're an heiress, and I—"

"We'll have all the way back to Mars to argue *that* out," she told him. "And I have an idea you'll lose."

Kenniston had the same idea.